2" Hot Cake and Sausage"

Part

The grandchildren are joining the Army.

"Mission", In The Hunt For " Paw Paw"

After the cruise, we finally made it home safe with our grandchildren, but only one came back with us, the others bad ass took a cab.

When they got to the house I said," where is the driver, and one of their bad ass said,' do you understand old ass man, we took the cab, with a gangster look on his face.

So I said," did you killed him, and the other said,' pull your shirt off, we know you are trying to incriminate us old man, so I said, what the

hell, how did you know I had a wire on me, and he said to me, 'do you want to visit him, wherever he is, so I said,' you all some bad ass kids," you all are going to the army.

So Kim and I went into the house, finally got off our feet, while these fools was plotting up something.

About that time, I ask Kim to turn the T.V. on, and there was a man, found tied up to a tree, with no clothing on, and someone put a chicken up his ass.

So I got online to find out, how to make a pill to enhance their size to a midget and make them talk like peoples. And there it was, under a protective website call terrorism.

So I waited until they all went to sleep, and I gave them the pill. The very next morning, when they awoke, my wife went to get them for breakfast and heard her said," Donald you need to come and see this" when I got there, these boys had a mustache and the girl had a mohawk.

So Kim went in the room and said," are GG babies ready to eat" and one of them said to her," look old woman, your food suck, and do something with your hair", now

I thought she was going to knock their ass out, but she acted like she did not hear a thing, and she left the room.

Now I told them," I gave your crazy ass some pill that will make you strong, and make you says words that we could understand"

and one of them said ,"I hope your old ass die" and I said "I can here you' 'and the other said," nigger we know".

Meanwhile, my computer was getting hack by the Tree topper, a gangster Villain overseas, the my computer crash.

So later that day, I took them to the recruiter to sign their ass up. When we got in the car, one of them said,"PaPa", your dum ass" and I said," what the hell you

want" she said," why didn't your ass fight for our country" and I said," I was to business smoking crack, and be quiet.

As we park the car, one of them slap me in 'the back of my head and they all got out, with that Army strong look on their face. Now I am feeling proud of my grandchildren, so I took them in and sign them up.

When we got inside, as I look through the window, and I saw a black SUV, park across the street, but I paid it no mind because the recruiter was about to talk to them.

So he ask them," why do you want to serve your country" and one of them said," none of your dam business" so I said to him," you see, they are Army strong.

So he ask them, can you kill your enemy" and one of the girls said," PaPa want die", he just multiply.

So he said on this day, you all have just joined the army. He told them where they will be station, and ask me do you have any last words, and I said, 'do they have to come back, and you can put there bad ass on the front line."

So they left for basic training. When I saw them off, I went outside to get in my car, and then I felt a bump in the back of my head, and to sleep I went.

I can remember for a breath minute, I was in a SUV, then to sleep I went back. I can remember, I was in a boat, and to sleep I went. I can remember, for a minute, I was on a plain, and to sleep I went I went.

When I awakening, there was a fool that look like a terrorist, and came up to me and said," where is the pill you made for your grandchildren", and I said," you kidnapped the wrong mother fucker" and he told me,"

well if you feel that way" and I said "I do", them he look over at his bitch boy and said," this nigger will lose a part of his body until he talk", so I said," ok I will tell you where my mother is, but don't take the dick."

It has been two weeks past, and my blonde ass wife just decided to call the police for a mission person. So went down to the police station, and said,' my husband is mission, and if he don't come back in so many days, can I get my 300 thousand life insurance policy", and they told her, 'hell no", did you kill him" and she said," every time I try to kill him, he just multiply.

So when she got through with reporting that I was missing, she started to think what I have been telling her on the cruise." Baby these dam kids are bad as hell, and they can talk", so she called the army to talk to them.

Now about this time, my grandchildren, these nuts have just finish basic training, and they all became Sgt. In the Army. Now before I go any further,

The pill I gave then, some how integrated in their DNA, meaning, these fool with be this way, always be like super nuts, now back to the story.

Now my wife was born a blonde, meaning that she just see things one way, and easy to judge," love you Kim when you read this book, but she went to the Base and sermon our grandchildren, and told them that Paw Paw has been kidnap, and here they come.

Meanwhile, while being interrogated, they are about to fuck me up, but I got to thinking, being a crackhead at one time, I can make this situation like court, so I ask them for a choir robe.

To my unbelief, these ass had one, and I told them to find a Bible, and I said really," you know your dumb ass don't have that over here, and they said," you fucking American", and knowing these fools want my pill, I said," you want the fuck pill, and court begin.

Now, I will not give out my grandchildren names, but I have to put a code name on them. So Kaya, o my bag, the oldest grandchild is level one, and the rest follow, ok.

Now by this time, my wife was talking to them, like she do not realize, this is the first time her ass had a conversation, never, but they came in, and they gave her the best hug ever.

So she told them that she think that your grandfather has been kidnapped by those folk with their face cover up, and the government will not help him, so I am asking you all to find his dumb ass, with no tears in her eyes.

While she had her head down, faking like she is crying, one of those fool said," you should be happy as hell "GG", and we know that you got an insurance policy.

Now about this time, level two said," I think GG want us to risk our precious life to save his gray ass, but she is on a difference planted.

Now "GG" fake cry got louder and the grandchildren was trying to fight back their tears, and then level four said," ok GG, for you, and only you we will try to bring that old piece of shit home, and she gave them all a hug, and some gum for there stinky breath.

While "GG" was leaving, there was a car following her, but she did notice, and keep on home.

Now while I was in a fuck up situation, I know in my heart, I am fuck up, so court has just began.

I look over, and there was a white man there with shit over his face, and I said," what is your name sir" and he said," why you pick the white guy" and I said,"

because your dumb ass think you are from here", this nigger said,". fuck the police", but he is my D.A.

So Abub, the leader got his defendant and we proceeded with trail. Now I said," white dude," call of first witness and he call the computer asshole, and he set down. I ask him, " where are you from Mr.," and this dump ass said," a boo boo, Rick James, whoop there it is", and I said," fuck this is going to be a long court, and I told his ass to get off my stand.

Two days past, they finally got together and they were discussing that should they let my ass die fast, are slow.

So level one said," do you all think he is our Paw Paw, and GG got a thing for black men, but whatever, let take a close picture who his dumb ass is.

So level three said," he make "GG" speak another language when they in their room at night, day, and sometimes,"yep", and she will be happy for two days, so I vote to save his old ass.

All in favorite of bring the old fart home, raise your hand" and about that time, no one raise their hand. So level three said again," all in favorite of making "GG" happy, raise your hand" and I was in the "mission."

So while I was having court, the terrorist got up and said," if you don't set your old ass down, talking about some dam court, I will blow your dam head off" and I said," the court is over.

I was beating for two days, and they left a pig in the room with me, to suffer from the smell. I had lost about forty pound when I first arrive.

When I got up, and look out of the window, I was on an Island, full of jungles and their soldiers.

When my wife got home from a exhausted day, he open up her door on the car, and about that time, a man came up to her and said," are you Ms. Quinney", and she said to him," me don't speech English" and he said to her," what you just said, is English."

So she told him," what in the hell you want" and

He said," your husband is kidnapped by a terrorist group, just want some information concerning some pill he made, do you know anything about that.

She looked at him and said," like that pill that make you stronger, the pill that can cause your IQ to be off the charts, so what, hit you up, and got you man."

And the man in the dark clothing said," what the hell you just said, but yep, those pill, and she told him," hell to the mother fucking no, and he left.

Now she is feeling very said about the events that is taking place in her life, so she took her a hot bath, and when she got into the bedroom, she got on her knees and said.

"Watch over my grandchildren and bring them home safe", and I would love to see my husband come home, because he still owe me child support."

She got up, pull off her slipper, and cut off her light, and all of a sudden, she cut on a powerful machine.

Now my grandchildren got around the table, they was all dress in the cool as tactical soldiers uniform, and they had a picture of me, smoking crack.

These bad ass kids, was bad ass soldiers, and I am too proud of them. Meanwhile, in the jungles, their leader, was talking to his sgt, and I heard him say," we must find those pill, because our army need them, so whatever you want to do with him, and make him shit himself.

In my mine, I said," what the fuck, so I got down on my knees, and kill the pig, just in case when I shit, it will take longer and maybe I can stank them out.

Now the day of the "mission", so level 1 and 2, was coming for me in the air, and level 3,4, by ship, and 5,6 it to dam high to know which way to go, so they got a submarine and they got together and said these words.

This day, is our Independent Day, and one said," no dumb ass, this day we try to kill are save Paw, Paw, and they left for the Island.

Now while 1 and 2, while flying in the air, they were talking some of the good time they had with me. "Do you remember, when we were on the cruise, and the other one said," how can I forget that shit."

When Paw Paw, left us in the helicopter and we took that bird on a awesome flight, and the other one said, 'we shot the shit of those boats, and the other one said," that was some cool shit."

So they look at one another and said," this is for you "GG", and they pull out a joint, turn on some "Sweet home Alabama" and got high as hell, while flying a jet.

3 and 4, was on the ship, thinking back on some good time, and said, 'do you remember when you was being a stripper, and by the way, what did you do with all that money"

So 4 said," first I bought me a lake house, but don't tell Paw

Paw ass, and I think Paw Paw, stole the rest of it for crack are something, because on the cruise, when he caught me, shaking it like a salt shaker, I told him, I will give him some.

They said to one another, I kind of miss his ass, and the other one said, I don't but, he isn't that bad and let get this mission accomplish, and they speed off.

Now 5 and six was thinking about some good time they had with Paw Paw, so one of them said,"

Do you remember, in his first stand up comedy book, not the second book, When his dumb ass took us to the park, and I call "GG" on his ass for cheating on her ass" and the other one said," we fuck Roscoe up, the dog," and "GG" knock dust out his ass.

You know what I miss the most about his old ass, while looking at each other, and about that time, 4 said," that a crazy old man, so let go get his dumb ass, and they got back on the mission.

So while the other two saw the camp that they had me in, they all went to the rally point to discuss the plan in saving me.

Meanwhile, when they made me clean up the dead pig, here come a man, with something look like a jumping cable, with a battery hock with it.

So I said," who car you going to jump off today" and he said you are to fucking funny" and I said, no mother fucker, I was for real, and about that time, two jump me from behind, and tired me up to a chair, put a cover over my head, and put my feet in a bucket of water, and lite my ass up.

Then for a brief moment, I saw Jimmy Carter, Jessie Jackson and Jesus at the same time.

After coming from the first shock, and said " can we have

court" and they pop me again and I passed out.

Now when they got their plan to rescue me, 4 said, 'do they have a strip club on the Island, and the others told him to shut the hell up and let's go.

Now for a minute, I will describe to moment. It was a

cold breezy morning, and you can hear the ocean, waving back and forward.

They went with, three one way, and two and a half the other way, now let get back.

One company went east and they came upon two guard, and they need to get in the gate, before the other company get to their position.

So 2 came up with a plain. While the guard was standing at the gate, all of a sudden, they had heard what appear

to be drums, beating from the woods, then here 2 come.

Jump out of the bushes and start dancing to the beat, " I am In Love With a Stripper' 'and this fool, jump on a tree with nothing on but some dam leaves, and drop it like it was to dam hot.

The guard came over and start making it rain, and about that time, the other jump out the woods, and cut their heads off, and went to the other side of the gate.

Now they call the others and they said," when you see a red

flair in the air, come with guns wide open."

The other company had made it to the building in which I was in and while looking at these fools from my window, they was outside the window, gambling on, what will they cut off my ass first.

I wanted to know, before they kill me, what about these pill they wanted. So I ask that motherfucker," why are you doing this to me because of some pill, and he said," do you remember the site that you got information from" and I said yes", you intercepted to direction from us, and you computer destroyed all information.

So I said," why in the hell, you guys fuck the computer up, but I will tell you what you wanted to know, but don't put that up my ass again.

Then the signal came, and these little bad ass grandchildren started shooting up the camp, even tried to shot me, but they rescue me.

When we all got back on the plain, they were discussing

something, but I could not understand what the hell they was saying, but

heard the world drop.

So I," what the hell you all are talking about, and about that time, the hatch door open, theses mother fucker drop my dumb ass off in the oceans and push a dam blow up boat and said," see your ass when we see you, the end.

PS. It was early morning, when they call "GG", to let her know, we rescue that nigger, but we dump his ass in the ocean but he be home soon, hell he smoke crack for 22 years.

When she got to her car to go to work, someone put a hood over her head." The end 2"

Hot Cake And Sausage

Part 3

" GG" is Missing"

When my grandchildren finally rescued me, there dumb ass drop me in the ocean, with a baby float. So when I finally came home, there was a note that said," If you want to see this bitch again, we need the Pills", but they don't know, it was destroyed with my computer.

So I had to call these bad ass kids. So I made the call, 'hello" one said, and " we see your dumb ass made it old man" and I said to his ass,"

www.ingramcontent.com/pod-product-compliance
Lightning Source LLC
Chambersburg PA
CBHW061222180526
45170CB00003B/1120